SCIENCE ADVENTURERS

VOLCANOLOGISTS

BY MARTHA LONDON

CONTENT CONSULTANT

Thomas Giachetti
Assistant Professor
Department of Earth Sciences
University of Oregon

Essential Library

An Imprint of Abdo Publishing
abdobooks.com

ABDOBOOKS.COM

Published by Abdo Publishing, a division of ABDO, PO Box 398166, Minneapolis, Minnesota 55439.

Printed in the United States of America, North Mankato, Minnesota.
092019
012020

THIS BOOK CONTAINS RECYCLED MATERIALS

Cover Photo: Krafft/Explorer/Science Source (front); Ksenia Ivashkevich/Shutterstock Images (back)
Interior Photos: Nina Janesikova/Shutterstock Images, 4–5; Jeremy Bishop/Science Source, 8; Pietro Pazzi/iStockphoto, 10–11, 52–53; DEA/Biblioteca Ambrosiana/De Agostini/Getty Images, 13; Shutterstock Images, 14, 69; Jae C. Hong/AP Images, 16–17; iStockphoto, 20–21, 29, 39, 42–43, 86–87; U.S. Geological Survey/Science Source, 24–25; Inger Eriksen/iStockphoto, 26–27; Benny Marty/Shutterstock Images, 30–31; Lena Klimkeit/picture alliance/Getty Images, 34; R. Stelmach/iStockphoto, 41; Mark A. Schneider/Science Source, 46, 90; Justin Reznick/iStockphoto, 48–49; Jessica Wilson/NASA/Science Source, 57; Mauro Fermariello/Science Source, 59; Kyodo/AP Images, 63; Xavier Arnau/iStockphoto, 64–65; Georg Gerster/Science Source, 66–67; Frederick R. McConnaughey/Science Source, 73; Doug Perrine/Alamy, 75; Victor Ivin/Shutterstock Images, 77; Tatiana Dyuvbanova/Shutterstock Images, 78–79; NASA/GSFC/METI/ERSDAC/JAROS, and U.S./Japan ASTER Science Team, 81; Marcel Strelow/Shutterstock Images, 88–89; Mario Ziebart/Shutterstock Images, 94–95; Brian van der Brug/Los Angeles Times/Getty Images, 97

Editor: Melissa York
Series Designer: Laura Graphenteen

LIBRARY OF CONGRESS CONTROL NUMBER: 2019941974

PUBLISHER'S CATALOGING-IN-PUBLICATION DATA

Names: London, Martha, author.
Title: Volcanologists / by Martha London
Description: Minneapolis, Minnesota : Abdo Publishing, 2020 | Series: Science adventurers | Includes online resources and index.
Identifiers: ISBN 9781532190360 (lib. bdg.) | ISBN 9781532176210 (ebook)
Subjects: LCSH: Volcanologists--Juvenile literature. | Geophysics--Juvenile literature. | Scientists--Juvenile literature. | Discovery and exploration--Juvenile literature. | Adventure and adventurers--Juvenile literature.
Classification: DDC 551.2107--dc23

CONTENTS

CHAPTER ONE

LIFE IN A LAVA LAKE

Rain had been falling for days as the team prepared to descend into the volcano's crater. The researchers and support crew stayed in their tents and waited. When the weather cleared, geobiologist Jeffrey Marlow suited up in rock climbing gear. Expedition specialist Chris Horsley joined him. As a rope climbing expert, Horsley's job was to keep Marlow safe within the crater. The week before the scientists arrived, Horsley had set up a series of anchors on the inside walls of the volcano. A system of pulleys and locks would attach the climbers' waist harnesses to nylon ropes. The rest of the team would stay at the top of the volcano.

MOUNT MARUM

The nation of Vanuatu is a series of volcanic islands in the Pacific Ocean near Australia. One of the islands is Ambrym. Vanuatu is difficult for outsiders

The lava lake in Mount Marum on Vanuatu is treacherous to reach.

to reach, and Ambrym is especially remote. Ambrym is a shield volcano. These volcanoes have gently sloping sides, but they are massive. They have a very wide base and cover a lot of area.

Ambrym has a huge caldera, which is a depression at the top of the volcano. A caldera usually forms after a large explosive eruption. After magma leaves the chamber, the rock above the chamber sinks down. Ambrym has two cones within its caldera, called Marum and Benbow. Cones are shaped like hills and build up around vents where lava escapes. Both of these cones have craters. Lava sits within these craters and erupts frequently.

Volcanologists from all over the world travel to Vanuatu to study Mount Marum, which has one of the world's few accessible lava lakes. Inside Mount Marum's crater is a boiling lake of molten rock. It's a 1,200-foot (400 m) drop from the top of the crater to the lava lake. Researchers must rappel down the wall of the crater.

Mount Marum is an explosive volcano. Its eruptions spew lava and toxic gas into the air. Sulfur gas mixes with the air and creates acid rain. This rain can harm plants and even eat through metal.

NAMED FOR THE GOD OF FIRE

Volcanology has its roots in Greek and Roman mythology. The word comes from the Roman god of fire, Vulcan, who was closely associated with volcanoes. According to the Romans, Vulcan lived beneath a volcano on the island of Vulcano, which is located north of Sicily in Italy. The Romans thought that the glow of eruptions at Vulcano was from Vulcan's forges as he worked beneath Earth. Records exist of a yearly festival in Vulcan's honor called the Vulcanalia. At the festival, people gave offerings to ask Vulcan to protect them from fire.

Scientists are searching for answers about what life could survive here, if any. They want to find out at what point conditions are too extreme for organisms to thrive. Next to the lake, the researchers hope to find out whether life can survive so close to lava.

THE DESCENT

Looking over the edge of the volcano, Marlow was excited and nervous. On camera, he said, "I study life in extreme environments, but this is certainly the most extreme environment I've been to. . . . It would be pretty remarkable to me if there's actually life down inside the crater."[1] The temperatures inside Marum's crater are very hot. On the day of the descent, Marlow carried a pack full of equipment. He had a temperature gauge to take readings at the bottom of the crater, as well as sample jars for collecting soil and rock samples from near the lake. These samples would go back to a lab where researchers would analyze the rocks for signs of microorganisms.

He also carried a thermal suit. Marlow would need protection from the heat when he reached the lava lake. The head-to-toe protective thermal suit went over his clothes. The bulky suit was made out of a silver reflective fabric and had thick gloves. Moving in the suit wasn't easy. The suit's helmet had a special shield that allowed Marlow to see while still protecting his face, but the gloves made it hard to handle small samples.

LAVA VS. MAGMA: WHAT'S THE DIFFERENCE?

Lava and magma are physically and chemically the same material. However, magma only exists under the surface of Earth's crust. There, high heat melts rock. In times of eruption, the magma moves toward the surface. Once magma hits the air, it is called lava. Lava cools and hardens, adding to Earth's crust.

Volcanologists' suits allow them to get close to lava and volcanic eruptions, but the work is still dangerous.

Marlow and Horsley leaned over the crater and stepped off the edge. When they came to the first ledge, they unhooked from the ropes and crossed slippery rocks to the next anchor. The area was full of loose gravel. If either man slipped, nothing would stop him from falling.

When the pair reached the boulder-filled field leading to the lava lake, Horsley stood back. Horsley's job was to make sure Marlow safely got to the bottom of the crater and back, but he wasn't collecting samples. Marlow stepped forward. Before he put on his thermal suit, he collected a few dirt samples from above the lake. Rain the day before had made the bare ground soft. Marlow drove a hollow cylinder into the mud, getting samples from the surface as well as from deeper layers of soil.

THE LAKE

Marlow got closer to the lava lake. Horsley stayed above to monitor the rock and weather conditions. Marlow put on the thermal suit. Gloves covered his hands, and boots protected his feet. Marlow carefully made his way as close as he could to the lava lake, but the helmet restricted his vision. A camera inside the suit filmed what Marlow saw. However, Marlow couldn't see to either side. He scanned back and forth to make sure each step he took was solid. A slip would be fatal.

Marlow stopped on a ledge with his back facing the lava lake. He hunched down as he held a pick similar to a small hammer. Loose rocks and gravel slipped and tumbled around him. Marlow hammered at a boulder, filling test tubes and containers with rock and soil samples. If the lab found microorganisms in these samples, it would give scientists a better understanding of how and where life might exist on other planets.

There are about 1,500 potentially active volcanoes in the world.[2]

PROTECTIVE GEAR

Most volcanologists don't spend a lot of time close to lava-filled craters. However, they still need to take precautions. The lava fields surrounding a volcano are full of jagged rocks. Volcanologists wear heavy boots, gloves, and hard hats in the field. The hard hats protect them from falling rock fragments. During periods of activity, volcanoes expel toxic gases, so volcanologists wear gas masks. Scientists only use silver thermal suits if they're going to be close to lava or the volcano's crater. The temperature of an active lava flow can rise to approximately 2,280 degrees Fahrenheit (1,250°C), depending on the type of rock.[3] The full-body suits protect the researchers from the high temperatures and harmful gases. But volcanologists try to get as much done as possible without the bulky suit because it makes movement difficult.

Before heading back to Horsley, Marlow took a temperature reading of the rocks. His thermometer read 150 degrees Fahrenheit (66°C).[4] Previous studies have established 250 degrees Fahrenheit (120°C) as the upper temperature limit for life.[5] With samples in hand, Marlow began the journey back to the top of the crater with Horsley. It was a successful trip. Marlow had collected valuable data. All that was left was to determine whether there was life in any of the samples.

THE RESULTS

Back at the camp, Marlow's colleague Jens Kallmeyer from the German Research Center for Geosciences prepared the samples for study. Kallmeyer used a fluorescent dye to stain the samples. This dye attaches to DNA, which is a chemical found in living organisms. If the samples glowed, Marlow and Kallmeyer would have proof of life.

Ambrym is surrounded by jungle with no roads.

> **"This is the frontline of astrobiology and the idea of looking for life beyond Earth."[6]**
>
> —*Volcanologist Jeffrey Marlow*

Under a microscope, Kallmeyer showed Marlow what was in the soil and rock samples. Microorganisms sat suspended and glowing on the microscope slide. The tiny single-celled organisms were proof of life in the crater of Marum's lava lake. If bacteria can live next to a lava lake, it's likely that life exists elsewhere in the universe. The conditions in a volcanic crater are extreme. But finding microorganisms there gives scientists a clearer idea of where life can survive.

VOLCANOLOGY

Volcanologists are geologists, physicists, and chemists who specialize in volcanoes. But they do more than visit active volcanoes. Studying the eruption itself is just one aspect of volcanology. Many volcanologists spend most of their time studying extinct and dormant volcanoes. The dried lava flows and ash fields contain many clues about how and why volcanoes erupt.

Scientists began developing systems to track volcanic eruptions in the 1800s. As technology improved, tools for studying volcanoes became more advanced. During this period, several scientists developed improved seismographs to measure even the smallest earthquakes. A seismograph's needle moves up and down while a drum covered with paper spins. If there is very little or no movement, the pen attached to the needle creates a straight line. If there's a lot of movement, the pen moves up and down in wide waves, mimicking the waves of the tremors.

Volcanology became even more important after destructive eruptions such as the 1980 Mount Saint Helens eruption in Washington State and the 1991 eruption of Mount Pinatubo in the Philippines. Large eruptions such as these provide scientists with a lot of data. But they also may cause lots of casualties and damage. Part of a volcanologist's job is to try to predict whether an eruption will occur and, if so, when it will happen and how long it will last.

People have tracked volcanic eruptions for thousands of years. For example, recorded eruptions of Mount Etna in Italy date back 3,500 years. These accounts help current volcanologists. A highly active volcano such as Mount Etna may have

Mount Etna erupted in 1843.

People have always been terrified and fascinated by volcanoes.

special factors that cause it to be more eruptive than other volcanoes.

People are drawn to the power of volcanoes. They are destructive. But they are also beautiful. Volcanic eruptions create fertile soil that helps plants thrive. Volcanoes are responsible for much of the life that exists on Earth, and scientists theorize that they are the source of our oceans. Volcanologists believe studying the forces that created our planet will help us better understand the conditions needed to support life on other planets in the universe.

THE DESTRUCTION OF POMPEII

Mount Vesuvius is located in the Bay of Naples in Italy. The volcano is at least 300,000 years old.[7] In 79 CE, Vesuvius erupted. The city of Pompeii was located at the base of the volcano. Many townspeople were able to escape. However, approximately 2,000 people decided to stay or were trapped. Waves of hot ash and toxic gas poured down the side of the volcano. After the eruption, Pompeii was buried under ten feet (3 m) of ash.[8] Pompeii wasn't rediscovered until the 1740s. Researchers found the city preserved almost completely intact from the day of the eruption.

The eruption was not spontaneous. Volcanologists believe that small earthquakes shook the ground underneath Mount Vesuvius for years before it erupted. Written evidence exists showing that people were aware of the tremors. However, because they were so common, people didn't believe there was a real threat. Vesuvius had a long period of dormancy before the 79 CE eruption. People believed they were safe.

CHAPTER TWO

DECADES OF DATA

In 1983, Kilauea started erupting on Hawaii's Big Island. The eruption finally stopped in the spring of 2018 after a very intense phase, when the crater floor and lava lake collapsed. Kilauea has offered volcanologists a rare look at an erupting volcano. The volcano continues to provide researchers with clues to how volcanoes work. These clues help volcanologists better predict what the future might hold for Kilauea and the Hawaiian Islands.

MAGMA'S PATH

The Hawaiian Volcano Observatory (HVO) monitors Kilauea's activity. Geologist Thomas Jaggar established the HVO in 1912. The HVO is a part of the United States Geological Survey (USGS). Its 25-person team monitors all of Hawaii's volcanoes and tracks earthquakes in the area. The HVO relies on researchers from many scientific disciplines to study the volcanic activity

Kilauea's long eruption ended spectacularly in the spring of 2018.

The Hawaiian Volcano Observatory had to move out of its building on the rim of Kilauea because of the volcanic activity in spring 2018. Unstable ground meant the researchers had not returned as of summer 2019.

on the island. In 2010, researchers began a study on Kilauea's magma chamber.

Kilauea's magma chamber is massive. The magma is very dense, and scientists hypothesized that the movement of the magma could increase or decrease the force of gravity. The force of gravity depends on mass, so gravity on Earth varies slightly from place to place because the planet's mass is not spread out evenly. As magma moves, its moving mass changes the nearby force of gravity. HVO scientists found a pattern in the gravitational force on Kilauea. They knew that a magma chamber sat below the volcano. The magma's churning movement through the chamber shifted enough mass to be measurable. If the scientists could detect where the magma was, they might be able to predict where and when another eruption would begin.

Continuous gravimeters measure gravitational force. In 2010, volcanologists set up two meters at the top of Kilauea. Before this experiment, most researchers didn't use gravimeters. They are expensive pieces of equipment. Michael Poland, an HVO geophysicist, commented, "The big users are oil and mining companies."[1] But the team hypothesized that if oil companies could measure oil deposits, scientists could measure the movement of magma. Poland went on to say, "Kilauea is the world's most active volcano. It's erupted almost continuously since 1983. It's a natural 'lab volcano'—a great place to try and study something like gravity measurements."[2]

One of the meters recorded measurements every ten seconds from 1.2 miles (2 km) northwest of the erupting fissure. Researchers placed the second meter just 500 feet (150 m) east of the fissure. It took recordings once every second.[3] The meters recorded changes in the force of gravity. As magma moved from one end of the chamber to another, the force of gravity changed. Soon, a pattern emerged.

The magma was moving in a predictable pattern. Gravity fluctuations happened once every 2.5 minutes. It wasn't the movement that was surprising to scientists—it was the regularity. Poland said the movement "points to the idea that there's probably a lot of things going on in volcanoes, glaciers, wherever you look, but we haven't developed the tools to detect these sorts of things."[4] Poland hopes gravimeters will help scientists predict future eruptions more accurately. A rapid change in gravity or its pattern may indicate an eruption is coming.

THE UNITED STATES GEOLOGICAL SURVEY

The United States Geological Survey (USGS) was founded in 1879. It formed as a way for the United States government to combine all earth science studies under a single organization. Today, the USGS provides information for the government regarding scientific studies, including volcanic research. Its research includes predicting eruptions of volcanoes around the United States.

KILAUEA'S PAST

Volcanologists use a technique called carbon dating to estimate the age of some rocks and other matter. Carbon-14 is a radioactive form of carbon. Over time, the radioactivity decreases. Measuring this decrease can accurately date objects up to about 60,000 years.

The oldest dated rocks from Kilauea are 23,000 years old, though volcanologists estimate Kilauea's first eruption happened between 300,000 and 600,000 years ago.[5]

Don Swanson is a volcanologist studying Kilauea's eruption history. He notes how advances in carbon dating have allowed scientists to get a clearer picture of the volcano's history. Swanson and his team spent a lot of time at Kilauea's summit. Kilauea's eruptions were occurring on the sides of the volcano. Fissures in the rift zone, an area in the volcano's side, allowed lava to escape. The summit was less active than the fissures.

At the summit, Swanson and his team worked to create a timeline of Kilauea's past. Each eruption has its own features. Near the summit, Swanson found evidence in ash deposits that Kilauea hasn't always been effusive, or in a flow state. Flow eruptions move slowly. The eruptions ooze and run like a stream.

Volcanoes and humans often come too close for comfort in Hawaii, where lava engulfs this highway.

Other eruptions are explosive. Explosive eruptions send rocks and ash clouds high into the atmosphere. Explosive eruptions may produce pyroclastic flows, which are hot clouds of ash and debris that hurtle down the volcano's slopes. Most of Kilauea's rocky slopes are formed from flows less than 1,000 years old. Seventy percent of the shield volcano's lava flows were deposited less than 600 years ago.[6] But older eruptions' ash and lava deposits show that Kilauea has been explosive in the past.

Scientists now theorize that more than half of Kilauea's eruptions have been explosive. Swanson has evidence to support not only that claim but also historical reports that an eruption in 1790 killed 400 people.[7] Reliable written accounts of the volcano's eruptions didn't start until 1823. Near the volcano's mouth, Swanson's team found a hardened pyroclastic flow deposit. In the flow were sets of footprints. Around the pyroclastic flow, hunks of hardened lava rocks were scattered on the ground. The geologists took samples from the flow and the rocks. Carbon dating the samples confirmed that the eruption happened in about 1790.

TWO TYPES OF LAVA

Lava doesn't all look the same. One type is fluid and smooth. When it dries, this lava looks like long ropes. Another type is more viscous, or thick and sticky. Larger chunks appear to be floating in the lava. Both of these types of lava are known by their Hawaiian names. The smooth, fluid lava is called pahoehoe. The viscous lava is called a'a.

Pele is the Hawaiian goddess of fire and volcanoes. Legends say she lives at the top of Kilauea.

THE END OF KILAUEA'S ERUPTIONS?

After more than 35 years of near-continuous eruption, people often asked scientists when Kilauea would stop erupting. Volcanologists and other researchers had a hard time answering that question. They had access to many tools. But they could not predict the dramatic end to the eruption.

In early April 2018, pressure beneath the cone increased. By the end of the month, huge clouds of ash plumed from the volcano's cone. The caldera began to collapse. Magma flowed to the lower East Rift Zone, where it erupted from the surface. Lava flowed out from the rift and destroyed homes below the volcano. It was the worst eruption in 200 years.[8]

For months, the lava lake continued to drain into the magma chambers. Collapse events were frequent. They caused the land to shake as if an earthquake were occurring. But the shaking was the walls of the caldera falling into the now-empty lava lake. By August, the summit of Pu'u O'o, one of Kilauea's cones, had

> "Kilauea will be a very different place when it reverts to an explosive period, the latest of which lasted for 300 years between about 1500 and 1800."[9]
>
> —*Don Swanson, USGS geologist on Kilauea*

PYROCLASTIC FLOWS

Pyroclastic flows are one of the most dangerous hazards associated with volcanic eruptions. These flows move as fast as 100 miles per hour (160 kmh). Pyroclastic flows aren't made of lava. The flow is a collection of gases, volcanic ash, and other particles that are heated to a very high temperature, about 1,300 degrees Fahrenheit (700°C).[10] The flow is very light, so it can move quickly. However, it's dense enough that it hugs the ground like a liquid instead of floating into the air like ash clouds. As the flow races down the side of the volcano, it burns everything in its path.

completely collapsed, and in September, the HVO noted that lava flows had stopped completely.

In December 2018, the HVO declared Kilauea's 35-year eruption over. But volcanologists say that Kilauea is not dead. Seismic events are still occurring beneath the volcano. Land deformation continues to happen, showing that magma still moves beneath the volcano. Kilauea will erupt again.

Scientists at Kilauea will continue to monitor earthquakes and gravitational forces while studying past eruptions. With time and improving technology, volcanologists are figuring out the secrets that Kilauea holds.

A geologist with the USGS photographing lava movement at Kilauea, June 25, 2018

CHAPTER THREE

SUPERVOLCANOES

The term *supervolcano* became popular in 2005. A documentary aired on the Discovery Channel about the volcano beneath Yellowstone National Park in the United States. A supervolcano releases lava with a volume of more than 240 cubic miles (1,000 cubic km).[1] An eruption of this magnitude would have global consequences, including significant cooling of Earth's surface and a long-lasting famine. There are 20 known supervolcanoes in the world.[2] In recent years, volcanologists have realized these volcanoes could erupt again.

HOW DOES A SUPERVOLCANO FORM?

Huge pockets of magma sit below Earth's surface. Sometimes these pockets of magma reach the surface in a huge eruption. These calderas can be many miles wide. Supervolcano calderas can fill with magma again. Scientists are

Brightly colored hot springs at Yellowstone National Park get much of their color from the heat-loving microorganisms that live in them.

trying to find a way to predict when and how these magma pockets fill, as well as how long it takes for pressure to build up enough to create a supereruption.

Supervolcanic eruptions are uncommon. Volcanologists estimate that a supereruption happens once every 100,000 years.[3] The last known supereruption occurred 75,000 years ago in Indonesia on Mount Toba. Most of Toba's ash scattered in the ocean. But the scale of the eruption was massive. Samples of marine ash deposits taken 4,000 miles (6,500 km) away showed a chemical composition that linked the material to Toba's eruption.[4] Following the eruption, Lake Toba formed inside the volcano's caldera.

Volcanologists used to think that a supereruption took thousands of years to prepare. They didn't believe that enough magma could rapidly gather under a supervolcano. They also thought it would take an earthquake to open a pathway deep into Earth. Studies in Yellowstone in the United States and Campi Flegrei in Italy show that a supervolcanic eruption could happen more quickly or with less warning.

MASS EXTINCTION

Over 200 million years ago, 90 percent of marine life and 75 percent of land animals died.[5] Scientists have long searched for explanations. A team of volcanologists traveled to Siberia in northern Asia. They transported hundreds of pounds of volcanic rock back to the United States. Crystals inside the volcanic rock gave the scientists a clearer idea of the timeline of a series of eruptions as Pangaea, an ancient supercontinent, broke apart. With the crystals confirming the theory, scientists knew for sure the eruptions came before the mass extinction and contributed to it.

Serene Lake Toba's round shape hints at its violent history.

STUDYING YELLOWSTONE

Yellowstone's last major eruption happened about 631,000 years ago.[6] People are starting to worry that it is due for another eruption. Yellowstone sits over an active hot spot, which is a place just under Earth's surface where unusually hot temperatures help rock in the mantle melt. Yellowstone's hot spot fuels the park's many geysers and hot springs. But if the magma that sits in the mantle moved into the chamber, Yellowstone could be closer to eruption than scientists previously thought.

Volcanologists used to believe that such a large amount of lava would take more than 1,000 years to collect in the chamber beneath the caldera.[7] However, research conducted by Arizona State University graduate student Hannah Shamloo revealed that Yellowstone's chamber may have filled in a matter of decades at the time of the last eruption.

Approximately one-half of the world's 1,000 geysers are found in Yellowstone National Park.

Scientists discovered Yellowstone's caldera in the 1960s—90 years after Yellowstone became a national park. The volcano's caldera is 44 miles (70 km) across.

HOT SPRINGS

Thermal springs, or hot springs, can vary in temperature. Some are boiling when they reach the surface. Others are at a temperature comfortable enough for people to bathe in. Water seeps through rocks and settles deep beneath Earth's surface. There it's warmed by the hot rocks surrounding it. As it warms, it expands and pushes up toward the surface. Some hot springs shoot out of the ground as geysers. Others form deep pools where a wide variety of algae and bacteria thrive.

Shamloo and her team spent several weeks at a rock formation called Lava Creek Tuff in Yellowstone in 2017. Tuff is a type of volcanic rock. The site is home to a fossilized ash deposit from Yellowstone's last eruption. Shamloo's team collected hundreds of samples of ash and rock. Occasionally, bison and bears crossed their path. The team took notes on their observations. Even hundreds of thousands of years later, the physical landscape could provide clues to Yellowstone's eruptive past.

Within the ash and rock samples were trace amounts of crystals. Each type of crystal tells a different story. At one point, the crystals were part of the molten rock beneath Earth's crust. Changes in the type or the concentration of crystals told Shamloo what happened just before the eruption. The crystals form rings over time, like a tree. Analyzing these rings informs scientists about the temperature, moisture, composition, and pressure changes within rocks over time. The outer ring tells scientists what happened just before the eruption.

The outer ring on the crystal samples revealed a rapid change in temperature. Shamloo and her faculty adviser, Christy Till, believed a large influx of magma into the chamber caused the temperature change. Shamloo's research showed that more data was needed to figure out how so much magma could move that quickly.

STUDYING CAMPI FLEGREI

Campi Flegrei means *burning fields* in Italian. Campi Flegrei's caldera sits beneath Naples, Italy. With a population of 1.5 million people, Naples is one of the most densely populated cities on Earth—and it's sitting on top of a volcano.[8] Most people know about Vesuvius in Italy. But even people in Naples forget they're living on top of a volcano. It has been so long since it last erupted. In the 2010s, Campi Flegrei was showing signs that new volcanic activity could happen soon.

Most of the volcano is underground or in the Bay of Naples in the Tyrrhenian Sea. Like most supervolcanoes, this volcano doesn't have one single cone. It's a complex collection of craters within the caldera. Besides the craters, the volcano includes vents

YELLOWSTONE'S HOT SPOT

Scientists Derek Schutt and Ken Dueker wanted to find out how big and hot the hot spot under Yellowstone National Park really was. They measured the temperature of the thermal mantle plume that fuels the hot spot. Schutt and Dueker found that compared to its surroundings, the hot spot averaged 90 to 360 degrees Fahrenheit (50 to 200°C) hotter.[9] While this is very hot, the scientists said that other hot spots, such as those in Hawaii, are much hotter. Dueker and Schutt believe it's possible the plume became disconnected from its source deep underground. However, this doesn't mean that the volcano is extinct. It still has the ability to erupt.

A geologist monitors satellite images of the Naples area for volcanic activity.

and geysers that release volcanic gases. After its last eruption in 1538, Campi Flegrei was quiet for 400 years. Then seismic activity and magmatic movement began again in the 1950s. Another intense period occurred from 1982 to 1984.

Today, scientists use satellite measurements and other tools to track the ground's movement. In 2017, they found the ground had risen 13 inches (33 cm) in ten years.[10] The

Vesuvius Observatory of Naples monitors volcanic activity in the Naples area. Computer screens and seismographs cover the office walls. Beeping noises come through the speakers as new data points show up on the screen. Volcanologists at the observatory say their job isn't just to track volcanic and seismic activity. They must also act as an educational and reassuring body for the public. The observatory has a public phone line where concerned citizens can call to ask questions.

The technology and scientific understanding doesn't exist for scientists to predict the next eruption. However, in the case of Campi Flegrei, the long-term measurements of volcanic activity are providing volcanologists with more information. Previously, scientists believed that after every upheaval event, the volcano settled back down to its previous levels. Looking at the graphs now, scientists at University College London (UCL) say there's a recognizable pattern.

Christopher Kilburn is the director of the Hazard Centre at UCL and the author of a 2017 study of Campi Flegrei's movements. He explained, "By studying how the ground

The Solfatara is a collection of mud baths and hot springs in Campi Flegrei's caldera. The baths have been in use for thousands of years.

WHY LIVE NEAR A VOLCANO?

There are many reasons why people live near active volcanoes. Volcanoes are rich in minerals. After an eruption, these minerals nourish the soil and make it excellent for farming crops. Volcanoes are also tourist destinations, providing beautiful scenery for residents and visitors. Once people settle in a place, it's difficult to convince them to move. If volcanoes have long periods of inactivity, people may be more apt to ignore the dangers of living near a volcano or inside a volcano's caldera.

the country. Unfortunately, the issue is more complicated than just drilling a hole in the ground and pumping water into the magma chamber.

Yellowstone and other national parks are protected from energy companies. The 1970 Geothermal Steam Act prevents geothermal companies from drilling in national parks. This protects the parks' landscapes. The act specifically names Yellowstone as a protected area.

Aside from hurting the landscape, other possible risks are attached to harnessing Yellowstone's power. Brian Wilcox was one of the scientists studying how people could use Yellowstone's energy. He admitted that the volcano would be a very efficient source of energy. However, drilling into the chamber could cause the volcano to erupt. In an interview with the British Broadcasting Corporation (BBC), he explained:

> If you drill into the top of the magma chamber and try and cool it from there, this would be very risky. This could make the cap over the magma chamber more brittle and prone to fracture. And you might trigger the release of harmful volatile gases in the magma at the top of the chamber, which would otherwise not be released.[13]

For now, Yellowstone will not power the United States, and the risk of a supereruption remains. But volcanologists will continue to study supervolcanoes such as Campi Flegrei and Yellowstone to better understand how these huge systems work.

Drill rigs disturb the environment around them.

Mud pot in Rotorua, New Zealand

WHY DO MUD POTS SMELL LIKE ROTTEN EGGS?

Geothermal features such as the mud pots, hot springs, and geysers at Yellowstone all have a distinctive smell. Many people compare the odor to rotten eggs. But there isn't anything wrong with the mud pots. The smell comes from hydrogen sulfide. Microorganisms live in and around the geothermal features. They eat sulfur, a yellow, chalky mineral. This creates sulfuric acid. The evaporating acid turns into hydrogen sulfide, which gives off the rotten-egg smell.

Despite the smell, these geothermal features have been used for centuries as a medicinal tool. Rotorua, New Zealand, is nicknamed the Sulfur City because of its thermal lakes and mud pots. Many of these areas have been turned into spas. Sulfur is a good treatment for skin irritations, including acne. The heat and other minerals in the pools can also ease arthritis pain.

CHAPTER FOUR

WHAT MAKES A VOLCANO EXPLODE?

Some volcanoes tend to explode more than ooze. Kilauea in Hawaii isn't very explosive. But some volcanoes send rocks, ash, and hardened lava into the sky. Debris can fly for miles before it lands. Volcanologists noticed that when some volcanoes erupted, the event was almost always explosive. But they weren't sure why. They began studying the volcanoes to understand the process better.

WHICH VOLCANOES EXPLODE?

The Cascade Mountain Range is located in the northwestern United States. Large volcanoes such as Mount Saint Helens, Mount Rainier, and Lassen Peak

Mount Saint Helens blew its top and formed a new crater in its 1980 eruption.

GLASS, PUMICE, SCORIAE, OR BOMBS?

Volcanic glass forms when lava is rapidly brought to a relatively low temperature. This causes the lava to harden quickly when it reaches the surface. If the decrease in temperature is rapid, crystals can't form.

Pumice, scoriae, and bombs are all projectiles from volcanoes. Pumice and scoriae are both fragments of lava containing lots of holes called vesicles. The difference between the two is in the number and size of the vesicles. Pumice comes from very explosive eruptions and usually has very small holes. Scoriae are coarser because the holes are generally larger. Scoriae form during less explosive eruptions. In both cases, gas bubbles get trapped in the lava.

Bombs are projectiles that are larger than 2.5 inches (6 cm) across. Generally, they are made from pieces of rock that broke off during the eruption's explosion. They don't explode when they hit the ground. But they can cause a lot of damage close to the crater because of their size.

are some of its tallest peaks. These volcanoes also have explosive eruptions. They don't have the slow flow of Kilauea. Researchers at the University of Oregon hypothesized that there must be something different about the magma in these volcanoes. Understanding what makes the volcanoes in the Cascade Range different could help volcanologists understand explosive eruptions around the world.

Cinder cone volcanoes surround Lassen Peak in California, located in the southern part of the Cascades. Cinder cones have steep sides and often grow out of the side of a larger volcanic mountain. Lava shoots out their tops like a fountain. Kristina J. Walowski was a doctoral student at the University of Oregon. In 2015, she led a five-member research team to the cinder cones in Lassen Volcanic National Park.

The team needed sturdy hiking boots. The terrain around Lassen Peak was filled with rocks and loose soil. In contrast to the green fields on the other side of the park, the trails leading to the cones were barren. Once they reached their destination, the team got to work. They collected samples of basalt, a type of rock that makes up most lava flows. Like most basalts, the rock around Lassen contained a lot of the green-colored mineral olivine. During their growth, olivine crystals may entrap little blobs of the surrounding magma. When they harden, these blobs become tiny glassy bubbles called melt inclusions. Walowski hoped to analyze which minerals were stuck within these bubbles.

Her team used equipment from labs in Washington, DC, and Oregon. Walowski analyzed some of her samples at the Center for Advanced Materials Characterization in Oregon (CAMCOR), a high-tech facility at the University of Oregon. One analyzer, the electron probe microanalyzer (EPMA), sends a beam of electrons at the rock sample. Within the rock sample are many elements and compounds. Each of these compounds has a different chemical makeup. When the electron beam hits the sample, each element and compound emits X-rays. Each compound's X-ray is like a fingerprint. The EPMA read these X-rays to determine which substances were in Walowski's olivine samples.

Lassen Volcanic National Park was established in 1916 shortly after Lassen Peak's last eruption in 1915. Researchers believe Lassen Peak first erupted about 825,000 years ago.[1]

MOUNT SAINT HELENS

Mount Saint Helens is an active volcano in Washington that often produces explosive eruptions. Its most famous eruption happened in 1980 after an earthquake hit the area. A major landslide followed by a blast knocked 1,300 feet (400 m) off the top of the volcano. Ash clouds covered the skies, scattering particles through the surrounding states. The eruption caused huge landslides and mudslides. Rushing flows of hot gases, ash, and other particles flattened forests and buried cars and cabins. Fifty-seven people died.[2]

Olivine crystals formed inside a basalt rock.

Walowski found that her samples contained a lot of water vapor. That means water must have been present at the time of the eruption. Researchers hypothesized that the presence of water helps create explosive eruptions along the Cascade Range.

HOMEMADE LAVA

Other scientists have studied the interactions between lava and water by actually combining them and observing the results. Lava can be difficult to obtain at the source and keep in its liquid state for study. So scientists made their own.

To make lava, researchers at the University at Buffalo in New York melted basaltic rock using a massive furnace. For four hours, the furnace heated the basalt to 2,400 degrees Fahrenheit (1,300°C). Once the basalt was in its liquid state, researchers poured the lava into insulated steel containers. The containers ranged in height from eight to eighteen inches (20 to 46 cm).[3] From there, researchers could add water to the samples. Researchers had to go to a test site to do their study because the components of their experiment were too dangerous to handle in a lab.

One such test site is about 40 miles (60 km) south of the University at Buffalo. The Geohazards Field Station allows scientists to conduct large-scale experiments that replicate real-life geological events. Volcanoes are unpredictable. To create a good experiment, scientists need to control some of the variables. In this case, researchers needed to control the height of the container and how fast water was injected. At a live volcano, this wouldn't be possible.

Scientists made ten-gallon (38 L) batches of lava. They poured the molten material into containers of varying shapes. After the lava was in the container, researchers injected water at varying speeds. Sometimes the reaction was spectacular. Large explosions occurred almost as soon as the water met the lava. But other times, the water evaporated quietly into the lava.

Researchers noticed that the container's shape made a difference. The speed of the water also affected the reaction. If the container was tall and narrow and the water was introduced quickly, the reaction was more explosive. Scientists aren't sure yet why this happened, but they have a theory. In 2018, researchers at the University at Buffalo published their initial findings in the *Journal of Geophysical Research: Solid Earth*.

Spectacular things can happen when lava meets water, as when lava flows into the ocean.

"Sometimes, when lava encounters water, you see huge, explosive activity. Other times, there is no explosion, and the lava may just cool down and form some interesting shapes. What we are doing is trying to learn about the conditions that cause the most violent reactions."[4]

—Ingo Sonder on the lava-water experiment

Ingo Sonder, the lead researcher on the project, has a background in physics and geosciences. The experiment didn't explore why container height and water speed affected explosiveness. The purpose of the study was to see what happened when water was added. It didn't address why the process occurred. But Sonder's background led him to guess a possible reason.

Sonder explained that when water is surrounded by a material that is much hotter, such as lava, a tiny portion of each water drop vaporizes. This water vapor becomes a protective shield that keeps the rest of the water from boiling and evaporating. This process is called the Leidenfrost Effect. When injected slowly into a shallow pool of lava, the water's vapor film stays intact, allowing the water to rise to the surface and escape as steam or evaporate into the lava itself. However, when water is introduced quickly, Sonder explains that it's possible for the vapor film to become unstable. When the vapor film disappears, the water boils, vaporizes, and expands rapidly. As the water vapor moves and expands through the column of lava, the lava must also expand. The lava moves outward and upward. The force builds as the lava continues to expand, creating a more explosive reaction. If Sonder is right, this could explain why the presence of water creates explosive eruptions in the Cascades. It may also help volcanologists predict the type of eruption to expect from dormant volcanoes that become active. Volcanologists can test olivine samples like the ones taken at Lassen Peak for the presence of water vapor. The amounts of water found in the samples would give scientists a better idea of how dormant volcanoes erupted when they were active in the past.

JÖKULHLAUPS

When a volcano erupts under a glacier, the hot magma melts the glacier, producing a massive amount of water. During this eruption, mud, rock, and chunks of ice may flood the surrounding area. Hot lava can melt so much ice that it produces five times its own volume in water.[5] Because this event happens frequently in Iceland, the phenomenon's Icelandic name, Jökulhlaup, describes the phenomenon all around the world.

THE RING OF FIRE

Stretching 25,000 miles (40,000 km), the Pacific Ring of Fire contains 75 percent of the world's known active volcanoes that don't lie deep under the ocean. From the coastlines of North and South America to the Aleutian Islands in Alaska and through Russia's Kamchatka peninsula, Japan, Indonesia, and New Zealand, the Ring of Fire has approximately 450 visible volcanoes.[1] Its shape is actually less like a ring and more like a horseshoe. Scientists have been studying the eruptions and earthquakes in the Ring of Fire since the 1960s. It's only in recent years that volcanologists are getting a better idea about what exactly is happening deep beneath Earth's surface to cause such explosive eruptions. It has to do with the movement of Earth's tectonic plates.

Many islands in the Ring of Fire are the tops of volcanoes. ▷

Earth's crust is divided into plates that are in constant slow motion. The Pacific Ring of Fire outlines the Pacific oceanic tectonic plate. The oceanic plate is denser than the continental plates surrounding it. When the oceanic plate meets a continental plate, the oceanic plate is pushed under the continental plate. This process is called subduction.

Subduction happens too slowly for humans to witness. However, it's easy to see the results of plate tectonics. Mountain ranges, earthquakes, and volcanoes are the product of this movement.

Subduction is also the process by which water enters Earth's mantle. The sediments that settle at the bottom of the ocean floor are pushed under the continental plate. Ocean sediments contain minerals with water in their chemical structure. This water is an important step in the creation of magma.

Super-heated solid rock forms much of the lower crust and upper mantle. The pressure from the plates keeps the rocks solid instead of melting them down into magma. As the oceanic plate shifts down, pressure and temperature increase. The water-rich minerals it contains destabilize, releasing their water. This added

TECTONIC PLATES AND EARTH'S LAYERS

Tectonic plates form Earth's crust. Earth is made up of three distinct layers: crust, mantle, and core. Tectonic plates are constantly on the move. Convergent plates crush together. Divergent plates spread apart. Both can form volcanoes. When tectonic plates converge, subduction can occur. One plate slips underneath the other. As the lower plate dives farther beneath the upper plate, it releases fluids under the increasing pressure and heat that help melt the mantle. The melted mantle can resurface as a volcanic eruption.

Many divergent tectonic plates appear at the bottom of the ocean. As the plates spread apart, the molten mantle rises up to fill the gap. Magma from the mantle erupts into the ocean, creating new oceanic crust. These volcanic eruptions form underwater mountain ranges, which are called the mid-oceanic ridges.

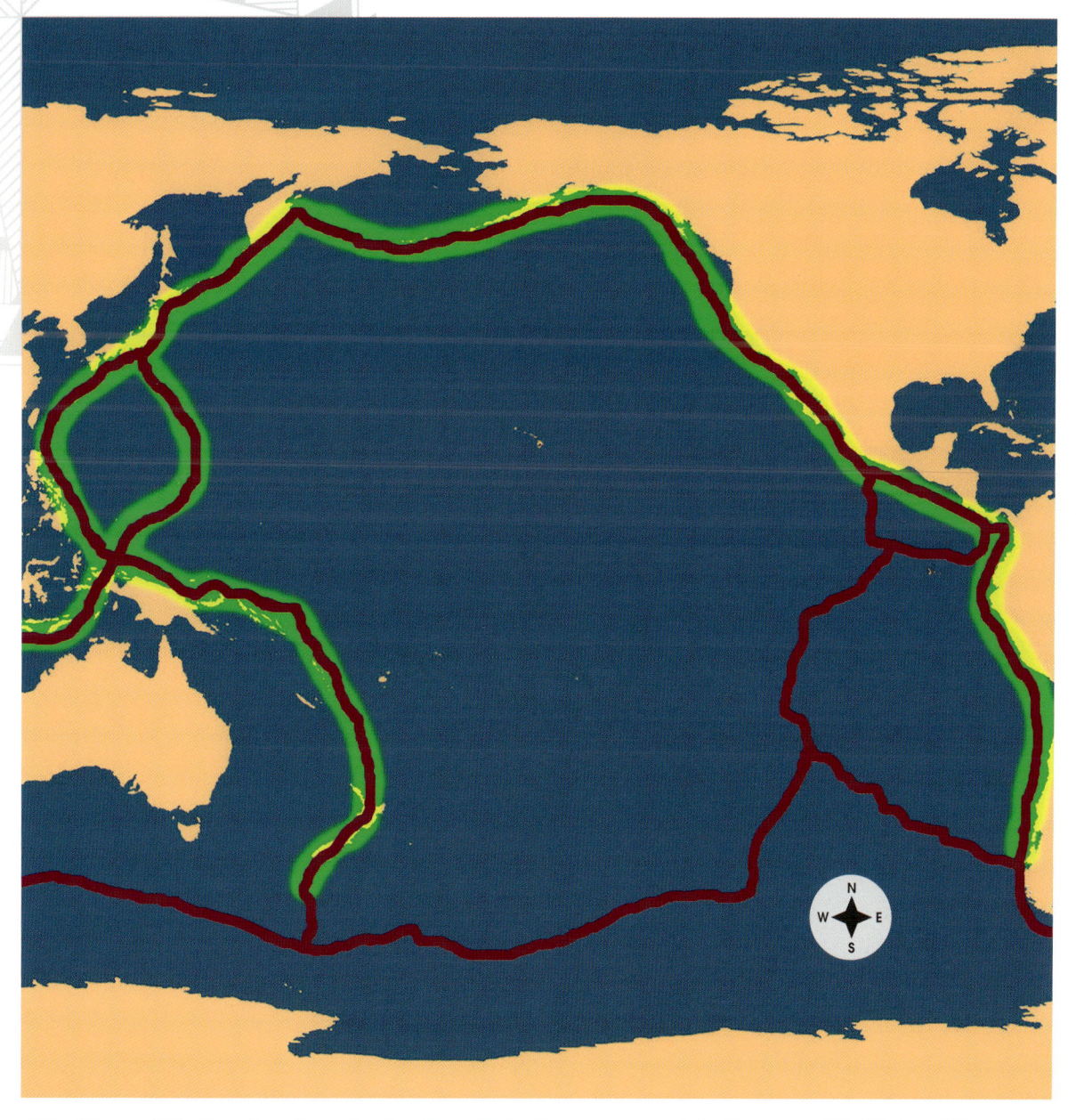

The Ring of Fire's volcanic and seismic activity happens most in the areas highlighted in green. The tectonic plates are outlined in red.

water lowers the melting point of the mantle rock. As the magma expands, it seeks out cracks in the crust and rises toward the surface.

But how do volcanologists know that it's water from minerals in the subducting plate and not a hidden source of water under Earth's crust or within the crust? To figure that out, USGS geologist Bill Evans headed to Lassen Peak. Along with lava flows, Lassen also has bubbling mud pots, hot springs, and fumaroles, a type of steaming vent. All of these features are powered by volcanic activity deep underground.

Evans sampled gases in a fumarole. Volcanologists all around the Ring of Fire have collected similar samples. This body of samples allows scientists from all over the world to compare data. Scientists are looking for patterns in their tests. They are also looking to see if the same tectonic patterns are happening in the entire Ring of Fire or just part of it.

The temperature of the steaming-hot fumarole hovers around 200 degrees Fahrenheit (93°C).[5] Wearing protective gloves, Evans used a long, hollow tube to collect a gas sample from deep in the fumarole, catching the gas as it rose through the tube in a clear glass container. Gases in the steam help scientists understand what's happening beneath the surface.

EARTHQUAKES

Ninety percent of the planet's earthquakes occur in the Ring of Fire. The constant movement of the tectonic plates sometimes jolts violently. Other times, people can't even feel the ground shaking. Seismographs monitor thousands of miles of land around the Ring of Fire. One seismic station in Alaska registers an average of 1,500 earthquakes each month.[6] The depth of the earthquakes appears to line up with the depth of the subducting Pacific plate.

Volcanologists around the world sample gases to study and monitor volcanic activity.

Analyzing the samples, Evans detected carbon-12. Carbon-12 is a specific type of carbon that comes only from living organisms. The presence of this substance tells scientists that at some point in the volcanic processes, the remains of organisms entered the system. Scientists believe these remains came from the ocean sediments of the Pacific plate. When organisms die, their remains settle on the ocean floor. Like the seawater trapped in the sediment, the organic matter also sinks beneath the continental plate.

WHAT DOES A VOLCANO SOUND LIKE?

The activity of the Ring of Fire allows scientists to brainstorm new ways of predicting eruptions. Because of the constant tectonic movement, some scientists are hoping to track a volcano's activity based on how it sounds.

Each volcano makes a different sound. Volcanologists call them voiceprints. Between 2015 and 2016, volcanologist Jeffrey Johnson studied the sounds made by Ecuador's Cotopaxi volcano. In an interview, Johnson said, "You can think of a crater as a musical instrument that's forming the tone as well as the reverberation of the sounds that we're recording."[7]

Johnson's studies of Cotopaxi in 2015 and Chile's Villarrica in 2013 weren't the first studies conducted on a volcano's music. Between 1992 and 2008, geophysicist Aurélien Dupont of the Pusan National University in South Korea and his team of researchers studied the sounds coming from Piton de la Fournaise volcano in the Indian Ocean. These sounds are at a frequency that the human ear can't hear. But if they were audible, "[they] would blow your eardrums," Johnson said.[8]

To record the sounds of the volcanoes, scientists use infrasound technology. *Infrasound* describes sounds that are too low for humans to hear. Scientists use condenser microphones and microbarometers to measure the sound frequencies coming from inside the volcano. Microbarometers measure tiny changes in air pressure. Condenser microphones are more sensitive than typical microphones. They are able to pick up sounds that people can't hear. These tools aren't placed on the volcano itself. In the case of Villarrica, sensors were placed five miles (8 km) from the volcano's cone.[9]

The sound waves have distinct shapes. Some, like those at Cotopaxi, look like the threads on a corkscrew. Johnson says these shapes give scientists a better idea of what the inside of the volcano looks like. According to Johnson's study, "The shape of these sound waves indicated that Cotopaxi's crater had become a 300-meter- [985-foot-] deep vertical shaft, creating infrasound frequencies similar to those of an organ pipe."[10]

As the magma levels change, so do the volcano's sounds. A longer, narrower tube creates a higher frequency. A wider, shallower pool has a lower frequency. If the volcano isn't in danger of erupting, the sound will be consistent from day to day. But as the magma and forces beneath the volcano change, the sounds associated with the volcano will correspondingly shift. Johnson is hopeful that careful monitoring of the voiceprints will allow volcanologists to predict eruptions. Having as much warning as possible before eruptions is important for keeping nearby communities safe.

WHAT CAN THE HUMAN EAR HEAR?

Sound travels in waves. These waves vibrate at a specific frequency. The faster the frequency, the higher the pitch. Sound waves are measured in hertz. One hertz is equal to one vibration per second. That is a very low sound. Humans have the ability to hear a range of frequencies. Most people can hear sounds between 20 and 20,000 hertz. In contrast, a bat can hear frequencies up to 200,000 hertz.[11] Infrasound such as the noise made by volcanoes is low. The frequency of volcano sound waves is slow, below 20 hertz and as low as 0.2 hertz.[12]

SCIENCE CONNECTION

TYPES OF EARTHQUAKES

Scientists define earthquakes based on where they form within Earth. The three types of earthquakes are shallow, intermediate, and deep. Shallow earthquakes occur near Earth's surface. Scientists define shallow earthquakes as occurring 0 to 40 miles (0 to 70 km) beneath Earth's surface. Because they are close to the surface, shallow earthquakes are capable of causing a lot of damage, even if they aren't very strong.

Intermediate and deep earthquakes are generally considered together in one category as any earthquake deeper than 40 miles (70 km). But deep earthquakes can be as far beneath the surface as 430 miles (700 km).[13] These earthquakes are caused by subducting plates slipping into Earth's mantle. The jagged edges of the tectonic plates catch on one another. As the tectonic plates continue to press together, energy is built up and then released in what is called a stick-slip displacement.

Scientists use seismographs to measure seismic waves caused by fault movements. They can analyze the waves' direction and intensity to determine the earthquake's depth and strength.

CHAPTER SIX

VOLCANOES AND OCEANS

Volcanoes have been present on Earth since early in the planet's history. They are constantly creating new land and participating in the water cycle. Scientists have debated for years where the water on Earth originated. Some scientists believed that meteorites could have left water when they hit Earth. Other scientists now say volcanoes are more likely the source.

THE METEORITE THEORY

During Earth's formation, the solar system was also taking shape. The planet was constantly bombarded with space rocks. For billions of years, debris

Scientists are working to discover the origin of Earth's water and what role volcanic activity may have played.

swirled around in space. During this turbulent time about 3.9 billion years ago, meteorites hit Earth, leaving deposits.[1] Remnants of these ancient impacts are still visible.

Upon impact, water vapor from the meteorites dissipated rapidly. Some scientists theorized that Earth's water came from these meteorites. However, based on data gathered from craters, it's more likely that any water molecules present in meteorites were blasted back out into space on impact.

Further research has shown that hydrogen, one of the components in water, is different on Earth and in space. Scientists call the form of hydrogen found in space *heavy hydrogen*. This, combined with evidence that rocks dating from more than 4 billion years ago interacted with water, leads many scientists to conclude that water already existed on the planet when the meteorites struck it.[2]

Meteor Crater in Arizona, created 50,000 years ago, shows what ancient impact craters looked like millions of years ago.

A BRIEF HISTORY OF THE WORLD

Earth formed approximately 4.5 billion years ago.[4] About 3.8 billion years ago, the first rain fell.[5] Since that time, Earth has had roughly the same amount of water. Shortly after the oceans filled, bacteria came on the scene. About one billion years later, algae that used a photosynthesis process emerged, creating oxygen. Through all of these changes, volcanoes formed, erupted, and eroded or sank back into Earth.

REVISING THEORIES

Scientists are still not in agreement about water's origin. However, one of the first studies to argue that water was already present when the meteorites hit was published in 1995. Peter Ulmer and Volkmar Trommsdorff worked at the Institute for Mineralogy and Petrography in Zurich, Switzerland. Their study showed that more water was found within the mantle than was introduced through subduction. When oceanic crust is being subducted in the mantle, some minerals break down due to the increase in pressure and temperature. When they break down, water is one of the molecules released. The water works its way upward through the region under the subducted crust. When the magma rises toward the surface, part of the released water is mixed in. When magma erupts, the water transforms into vapor and escapes into the atmosphere.

Ulmer and Trommsdorff's study focused on the area within Earth called the subduction zone. This area extends about 120 miles (200 km) beneath the surface.[3] However, new studies show that water may exist much deeper within the planet.

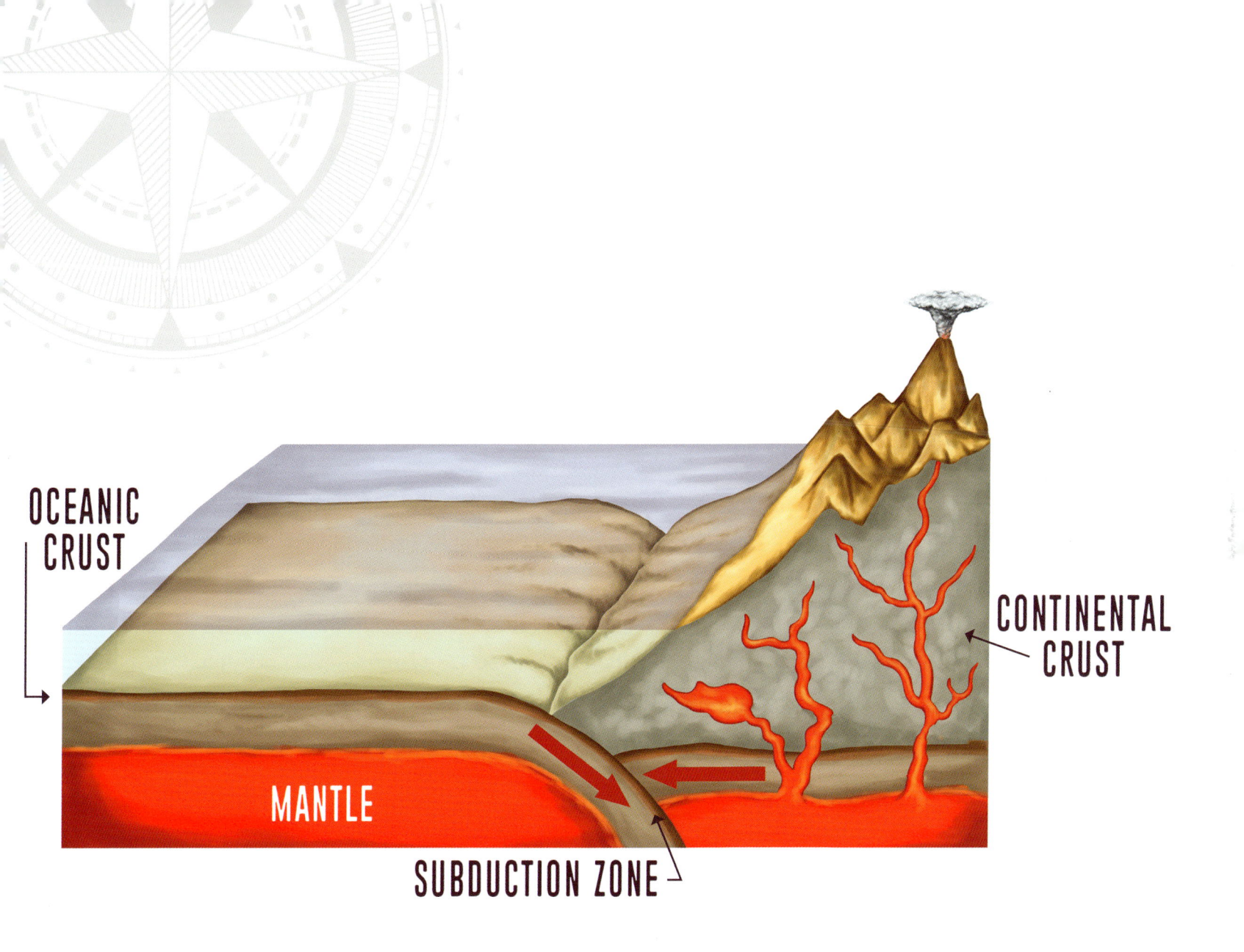

OCEANIC CRUST

MANTLE

SUBDUCTION ZONE

CONTINENTAL CRUST

Subduction happens in Earth's mantle.

Scientists from across the United States conducted a study in 2014. The team hypothesized that water was present far deeper than the upper mantle. Scientists came to this hypothesis based on the minerals found in the mantle transition zone, 250 to 410 miles (410 to 660 km) below the surface.[6] Minerals sink based on how dense they are. Lighter, less dense minerals settle near the surface, while denser material sinks until it settles in an area that is denser. This is similar to oil and water, as oil sits on top of water because it's less dense.

Scientists can't drill deep enough into Earth to test samples. Instead, the team of scientists ran computer simulations using numerical models. These models mimicked the conditions within the mantle. Because of the flowing currents in the mantle, materials are constantly on the move. The movement of materials downward is called downwelling. Based on the team's calculations and the minerals' reactions to pressure and heat, the scientists concluded that "the mantle transition zone . . . acts as a large reservoir of water."[7] That water is locked within minerals.

The release of water and gases through volcanic activity is called degassing. Degassing occurs through eruptions as well as other volcanic activity, including fumaroles, hot springs, and geysers.

A 2016 study involving scientists from Canada, the United States, Germany, Italy, and the United Kingdom built on these findings. They discovered that the partial melting of rock in the lower mantle is responsible for moving water up toward the surface. Most liquids are less dense than solids. The water within the magma helps lower the melting temperature of rocks. The solid mantle rock sinks, while the melted magma begins to rise.

Water, rock, and magma all work together. The water lowers the melting point of the mantle rock. But the process of melting the rock heats the water. The water expands. As it expands, it gets squeezed upward. Water can be absorbed into the magma itself and travel to the surface during an eruption.

These studies helped scientists rethink how Earth's water arrived at the surface. As the magma moves up through the mantle, it meets the crust. The magma moves until it finds cracks in the crust. When Earth was young, the crust was thin. Many volcanic eruptions occurred. With each eruption, water vapor was released into the atmosphere.

As Earth continued to cool, the water vapor in the atmosphere condensed. The condensed vapor formed clouds and fell to the ground as rain. Scientists estimate that it rained for thousands of years. The rainwater filled low-lying areas and eventually created oceans and seas. Dissolved sodium and chlorine that was present in rocks and the atmosphere at the time caused the oceans' salinity.

ACID RAIN

When volcanoes erupt, they emit sulfuric acid gas. These gas clouds spread through the atmosphere. Rainfall mixes with the sulfuric acid and can create acid rain. Acid rain isn't harmful to humans if it lands on skin. However, acid rain damages plants. It can also cause lead to leak into drinking water. The acid eats away at materials that contain lead, releasing the lead into water sources. Ingesting lead is dangerous for humans.

VOLCANOES IN THE OCEAN

Volcanologists estimate there are more than one million underwater volcanoes, including thousands that are extinct.[8] Some of these formations are huge. The ones that are more than 3,300 feet (1,000 m) tall are called seamounts. Smaller extinct volcanoes are called sea knolls. Many seamounts form in lines along the ocean floor. For example, in the Pacific Ocean, seamounts generally form in chains of ten to 100. One of the largest seamounts recorded, Great Meteor Tablemount, rises more than 13,000 feet (4,000 m) above the ocean floor.[9] Its top is underwater.

The ocean floor is the site of a wide array of volcanic activity. But tracking submarine eruptions is difficult. Scientists first discovered hydrothermal vents under the ocean surface in 1977. The scientific community knew volcanic activity was occurring, but the technology didn't exist for researchers to find these volcanoes. Many of these volcanoes are located near mid-ocean ridges. The average depth of submarine volcanoes is 8,500 feet (2,600 m) below the ocean surface. Submarine volcanoes are responsible for about 75 percent of the lava that erupts every year.[10]

MAPPING UNDERSEA VOLCANOES

In 2015, a group of Australian scientists were looking for lobster nursery grounds off the coast of Sydney, Australia. However, they stumbled on something much larger. The team's ship used sonar technology to map the ocean floor. Three miles (5 km) below the ocean surface were four volcanoes. The researchers estimated that they were 50 million years old. While the volcanoes are extinct now, the discovery was still exciting. The largest of the volcanoes rose 2,300 feet (700 m) off the floor. Its caldera was nearly one mile (1.5 km) across.[11]

Undersea volcanic eruptions are different than eruptions on land. Scientists believe the primary reason submarine eruptions act differently is due to the pressure water exerts on volcanoes on the ocean floor. Pressure produces flows that look different from the pahoehoe and a'a flows volcanologists study on land. Pillow lava, lobate flows, and lineated flows are common formations in undersea eruptions.

Pillow lava takes its name from the lava's shape when it hardens quickly in the cold ocean water. Pillow lava doesn't travel far. It creates rounded tubes. Often these tubes stack up, creating a lumpy chimney.

Lobate and lineated flows tend to cover more ground than pillow lava. A lobate flow resembles pillow lava. It has rounded edges. The top of the flow hardens quickly, but beneath the lava's crust, molten rock still flows. It bubbles forward along the ocean

Bubbling vents are one indication of underwater volcanic activity.

floor or the side of the volcano. Lineated flows form in the same way as lobate. However, instead of creating round lobes of lava, lineated flows harden with parallel lines. These lines indicate the direction of flow.

Volcanologists haven't often been able to watch these flows as they are happening. They rely on photos of the lava flows once they have hardened. Volcanologists use submersibles to study undersea volcanoes and vents. In the submersibles, volcanologists can film and take photographs of the lava flows. The patterns can tell scientists what kind of eruption occurred and how quickly it flowed and cooled. Some flows harden in parallel lines. Other flows are blockier. Scientists use the submersibles to take samples of the lava and sediment deposits. The samples can give the researchers a lot of information. For example, volcanologists use the sediment deposits on top of the flows to estimate the age of the eruptions.

SEA VENTS DON'T BOIL

Seawater in hydrothermal vents can be as hot as 700 degrees Fahrenheit (370°C).[12] But when the water comes out, it isn't boiling. That's because the pressure at the bottom of the ocean is so high. When the hot water from the vent meets the cold seawater outside, the water cools. As it cools, some minerals solidify and create vent chimneys. These chimneys can reach heights of 180 feet (55 m).[13]

Pillow lava hardens into round and lumpy shapes.

SCIENCE CONNECTION

ROBOTS AND VOLCANOES

Volcanologists use robots to explore hard-to-reach areas of volcanoes and volcanic eruptions. Motorized robots are able to access locations that scientists cannot, such as the bottom of the ocean. Robots can take samples of lava flows, rock deposits, hot water at a vent chimney, or gas from a volcanic plume. They then bring the samples back for analysis.

Robots used in scientific experiments are highly specialized. They have tools and abilities that help them navigate rough and remote terrain. At the bottom of the ocean, robots are remotely piloted. Scientists use controllers to steer them. Robots often have video cameras attached that send a livestream of what's happening beneath the surface. In this way, scientists can examine the location alongside the robot. Scientists can control what samples the robot takes. They can also make notes about the underwater volcano's physical appearance.

Once the robot is on the sea floor near the volcano, scientists maneuver the robot around. The robot's arms can extend and pick up large rocks. Some have vacuumlike attachments to take samples of sediment and seawater. When the robot returns to the surface, scientists can take the samples back to the lab for analysis.

CHAPTER SEVEN

PREDICTING ERUPTIONS

Scientists hope to one day predict eruptions accurately and with enough time to safely evacuate people from the area. However, predicting eruptions is very difficult. One of the challenges volcanologists face is that many volcanoes are dormant. Some haven't erupted for thousands of years. But most of these volcanoes have the ability to erupt.

One way to study the causes of an eruption is to examine what happens during and immediately after one occurs. Using new techniques and technologies, volcanologists are finding tiny clues that may help predict eruptions. These new technologies are also showing ways in which some volcanoes are physically connected and how small eruptions may be helpful in preventing supereruptions.

New technology such as drones helps scientists better monitor volcanic activity.

DORMANT, ACTIVE, OR EXTINCT?

Volcanoes' life spans can last millions of years. Because of this, volcanologists don't agree on what makes a volcano active or dormant. An extinct volcano is cut off from its source of magma and therefore can't erupt again. There is no clear distinction between active and dormant volcanoes. Many scientists say active volcanoes are those that have erupted in recorded history or just those that are currently restless and are expected to erupt again soon. Dormant volcanoes have erupted before and are expected to erupt again at some point. But they have been inactive for thousands of years.

SATELLITE TECHNOLOGY

The European Space Agency (ESA) launched the satellite *Sentinel* in 2014. The satellite is used to monitor the movement of land around volcanoes. Juliet Biggs of the University of Bristol in the United Kingdom used *Sentinel* in a volcano study. *Sentinel* captured images of more than 500 volcanoes around the world.[1] Over a series of months, each volcano had multiple images taken.

Sentinel was able to track tiny movements of the ground. The 2014 study found that 46 percent of volcanoes with ground deformation erupted. In comparison, 94 percent of volcanoes that didn't deform also didn't erupt.[2] Therefore, it's more likely that a volcano with ground movement will erupt than a volcano without ground movement.

Biggs was excited. She said, "The findings suggest that satellite radar is the perfect tool to identify volcanic unrest on a regional or global scale and target ground-based monitoring."[3] For locations with limited monitoring resources, *Sentinel* provides a global team for volcano monitoring.

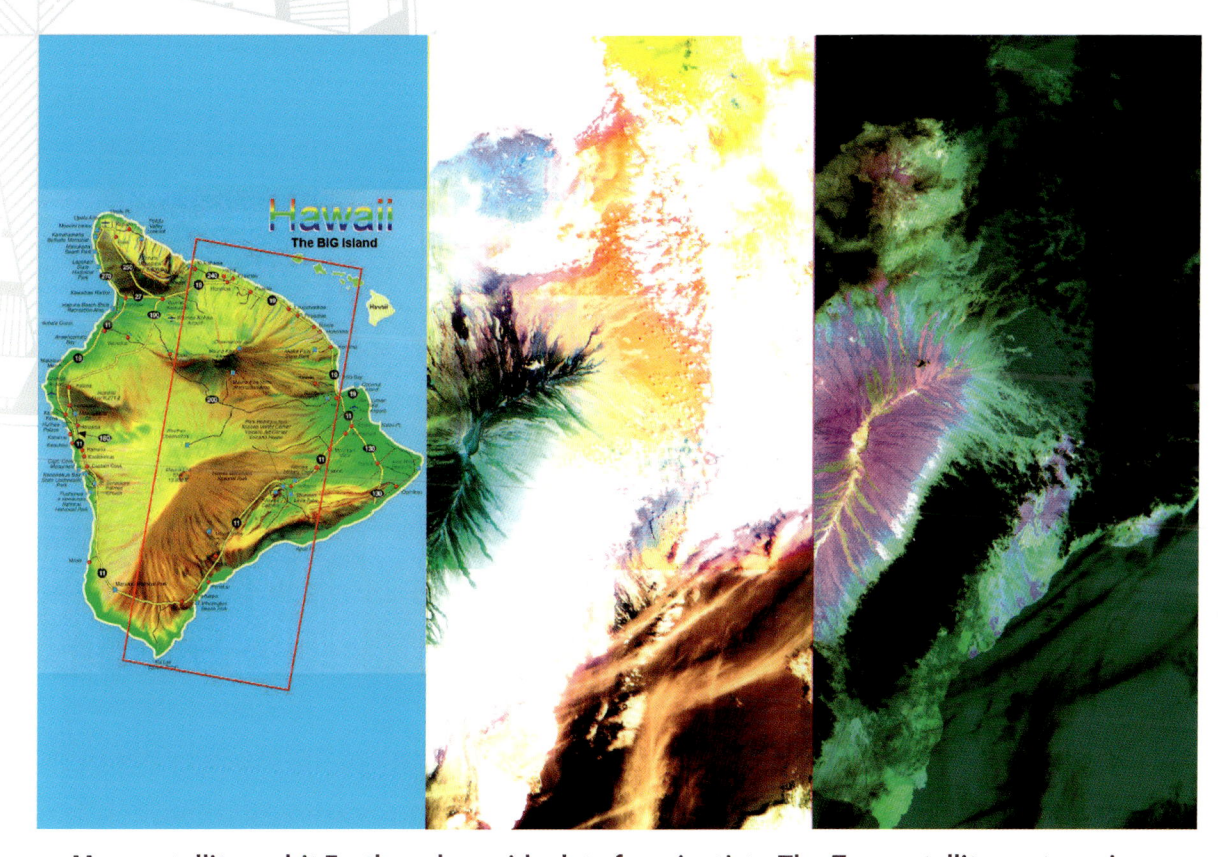

Many satellites orbit Earth and provide data for scientists. The *Terra* satellite captures images with visible light as well as thermal images showing heat, and together the data gives researchers a better insight into the area they are observing.

In 2017, Agung volcano in Bali erupted after a 50-year dormancy period. People living near Agung were evacuated before the eruption because monitors noted an increase in seismic activity. Agung has a neighbor volcano, Batur. In 1963, the last time Agung erupted, Batur erupted shortly after in a smaller eruption. The 1963 eruption killed 2,000 people, making it one of the deadliest eruptions in the 1900s.[4]

The University of Bristol teamed up with the Center for Volcanology and Geological Hazard Mitigation (CVGHM) in Indonesia to conduct a study to find out why Agung erupted. Biggs led the team. Using images from *Sentinel*, they saw a significant movement at the base of the volcano. The ground lifted nearly four inches (10 cm). But the images also showed the movement didn't happen near the volcano's summit. Instead, the uplift happened three miles (5 km) away.[5] That tells researchers that the magma beneath Agung isn't only moving upward. It's also moving side to side. Fabien Albino, a colleague of Biggs, said, "Our study provides the first geophysical evidence that Agung and Batur volcanoes may have a connected plumbing system. This has important implications for eruption forecasting and could explain the occurrence of simultaneous eruptions such as in 1963."[6]

TRAVELING THROUGH ASH

Volcanic eruptions send plumes of ash into the atmosphere. During the day, it's easy to see these clouds. However, at night, ash clouds are difficult or impossible to see. Ash is dangerous for airplanes. It clogs airplane engines and causes engine failure. Eruptions can disrupt air traffic for weeks as the ash scatters. Researchers are finding new ways to track and predict where ash plumes will travel so that more air travel is possible even during an eruption.

SLOW AND STEADY

Volcanologists are still working to determine the plumbing systems of volcanoes. Each volcano forms differently. After an eruption, where does the magma go? Does it settle deep underground, or does it move closer to Earth's surface? Determining this information could help scientists predict whether a volcano is in danger of erupting. It may also help scientists determine whether a supervolcano will always have supereruptions.

A research group from Vanderbilt University in Tennessee traveled to New Zealand in 2018. Volcanologist Guilherme Gualda led a group of students to the Taupo Volcanic Zone. The area experienced seven large eruptions between 350,000 and 240,000 years ago.[7] Gualda studies supereruptions. He wanted to know if magma pools changed with so many eruptions in a short time.

The team collected and analyzed pumice. Layers of pumice were found near the road. Others were on rocky outcrops. The team looked for the formation of crystals in the samples. There were few crystals in the pumice. This told the scientists that eruptions were more frequent. The magma was moving closer to Earth's surface.

Gualda thinks frequent eruptions are an important pressure release. When pressure built up, it was already close to the surface, and large amounts of magma weren't able to form. The volcano easily erupted, releasing the pressure from the magma pools and preventing large eruptions.

"As the [volcanic] system resets, the deposits [of magma] become shallower. The crust is getting warmer and weaker, so magma can lodge itself at shallower levels."[8]

—Guilherme Gualda, Associate Professor of Earth and Environmental Sciences, Vanderbilt University

What remains unclear is how long it took for these pools to form. Gualda admits that it could take thousands of years. However, he says, it's more likely that they formed in only hundreds of years. More research is needed to better understand the movement of magma from the mantle.

LEARNING FROM LAVA

Magma doesn't move at a constant speed. The shape of a volcano's underground plumbing may be a reason why. One way scientists can measure how quickly magma is moving is by examining the crystals that form in lava. These measurements help scientists better predict when an eruption will occur.

Two separate studies, one in 2017 and another in 2019, looked at crystal formation in the hardened lava flows of Mount Etna in Italy. The patterns of crystals present in lava tell scientists how quickly magma moved through the volcanic chambers during historical eruptions. A team of researchers from the University of New Hampshire (UNH) conducted the 2017 study. The 2019 study was headed by a team from the University of Queensland (UQ).

Sarah Miller led the UNH team. She said, "Some elements move rapidly and some more slowly, so there is a chemical record of events in those crystals that can help us determine

NEW ECOSYSTEMS

Lava tubes form after an eruption ends and the lava leaves the carved-out cavern. What's left is a long tunnel that creates a unique ecosystem. Many animals and plants are specially adapted to the environment of a lava tube. Insects there live in darkness and have no pigment. With time and erosion, roots from above the tube dangle from the ceiling. Rain soaks down through the plants and provides water and food to the animals in the tube. If the tube were to become active again, the entire ecosystem would be wiped away.

their journey."[9] Her team found that Mount Etna's magma source today and its ancient magma source are almost identical. They also found Mount Etna had several different storage chambers where magma began crystallizing. However, their research found that most of Mount Etna's magma wasn't stored very long before eruption.

Mount Etna's eruptions are often short and violent. These outbursts are called paroxysms.

Building off UNH's work, the UQ team measured trace elements in crystals. Some volcanic crystals have different chemical compositions. These variations are called sector zoning. Teresa Ubide and her team think these variations can be used to mark precise changes in the volcano's interior.

The UQ team used samples from historical lava flows. A laser beam sliced off a tiny portion of rock. This tiny slice went through a mass spectrometer. Mass spectrometers measure trace elements in a sample. They break apart atoms into ions by providing a tiny electric current. Then a magnetic field scatters the ions. Lighter materials scatter more. Heavy materials stay closer together.

The results from the mass spectrometer showed several things. The elements in the crystals were sensitive to what was happening in the volcano. The crystals were affected by tiny changes in the magma's movement, storage, cooling, and mixing. Ubide said, "Now we've discovered that [the crystals] not only record detailed magmatic histories and

eruption triggers, but might also provide information on the velocity of magma transport to the surface."[10]

Ubide says the information collected could have a big impact on how scientists predict future eruptions: "The new information on magma transport prior to past volcanic eruptions can provide context to help better respond to future monitoring signals, like seismic measurements from earthquakes."[11] With the new information, combined with seismic measurements, scientists may be able to better predict how quickly an eruption could happen after an earthquake.

Studying lava flows from the past gives scientists clues about potential future eruptions.

CHAPTER EIGHT

ERUPTIONS AND EARTH'S CLIMATE

Volcanic eruptions are destructive to the immediate area around them. They also cause changes to Earth's climate. Surface cooling and methane emissions contribute to the part eruptions play in global climate change.

ISLE OF SKYE

Approximately 56 million years ago, a large eruption occurred on the Isle of Skye in Scotland. But researchers weren't aware of it until 2019. Researchers knew that about 56 million years ago, there was a period of major warming around the globe. Temperatures increased by nine to 14 degrees Fahrenheit (5 to 8°C).[1] Scientists attributed the warming to a period of explosive

Ancient volcanic activity shaped the rugged landscape of the Isle of Skye in Scotland and also likely triggered climate change.

Pitchstone can be black, brown, gray, red, or green.

eruptions in the Northern Hemisphere. But most volcanologists assumed Scotland didn't have any volcanic activity during the time leading up to the period of warming.

A group of researchers from Sweden and the United Kingdom compared deposits of pitchstone, a type of glassy rock. Pitchstone forms when lava cools rapidly. It's not as shiny as obsidian, another type of volcanic glass, because pitchstone has more crystals that form within the rock. Therefore, it's a good rock for volcanologists and geologists to collect to compare crystal formations.

The group collected samples of pitchstone from rocky outcrops that were more than 18 miles (30 km) apart, on two islands in the Inner Hebrides.[2] Researchers used different methods to test whether the deposits were from the same eruption. One technique is to look at the samples under a microscope. Scientists look for identical formations. The other method is isotope geochemistry. Chemical elements have a certain number of protons and neutrons. Isotopes are variations in the number of neutrons that make the element lighter or heavier. For example, oxygen-16 has eight protons and eight neutrons. A common isotope of oxygen is oxygen-18. It has two extra neutrons. When pitchstone forms, multiple elements, including oxygen, create bonds and crystallize. Isotope geochemistry measures the isotopes of the pitchstone samples. The isotopes found in the samples act like a fingerprint of the eruption.

NATURAL GLASS

Pitchstone and obsidian are two types of natural glass formed from lava. The lava cools rapidly. Obsidian cools faster than pitchstone. This means that crystals don't have time to form. Pitchstone has more crystals than obsidian, but they are small. Scientists believe the presence of crystals is the reason pitchstone is less shiny than obsidian. People have used natural glass for thousands of years. Obsidian has been especially prized for its use in knives and arrowheads. Natural glass keeps its edge and is durable, even when it's carved very thin.

When the analyses came back, scientists found that the pitchstone samples were chemically identical. This finding told the scientists that there was a massive eruption around the same time as the period of warming. The data helped scientists reevaluate their previous theories about volcanic activity in the area. They now have evidence that the eruption on the Isle of Skye was a major factor in the warming of the planet 56 million years ago.

METHANE

When volcanoes erupt, one of the byproducts is carbon dioxide. Carbon dioxide emissions contribute to Earth's warming. But methane is a much more powerful factor. Peter Wynn is a glacial biogeochemist at the Lancaster Environment Center in the United Kingdom. He was one of the authors of a 2018 study from Lancaster University about glacial melting in Iceland. Regarding methane, Wynn explained, "Methane has a global warming potential 28 times that of carbon dioxide (CO_2). It is therefore important that we know about different sources of methane being released to the atmosphere and how they might change in the future."[3]

VOLCANOES IN ANTARCTICA

Antarctica has one of the highest concentrations of volcanoes on Earth. But this discovery is recent. Until 2017, volcanologists knew of 47 volcanoes on the continent. In 2017, a team of Scottish scientists found 91 more volcanoes under the ocean's surface on the continental boundary.[4] Scientists are now trying to figure out how active these volcanoes are. Ice melt in the polar regions has been increasing due to climate change. A volcanic eruption could speed up the process even more.

Katla is an active volcano in Iceland. It's covered in a large ice cap. Projecting from this ice cap is the Sólheimajökull glacier. Rebecca Burns studied the emissions of methane from the glacier while she completed her doctoral degree. To complete her study, Burns collected water samples. Some of the samples came from the lake fed by the glacier's meltwater. Burns also tested other samples from streams and soils for comparison.

Burns's tests found the highest concentration of methane in the samples taken closest to the river that flows into the lake. This led Burns to conclude that the glacier must hold a significant amount of methane. To confirm her observations, Burns used a mass spectrometer to locate the methane's source. Methane is a compound made of carbon and hydrogen atoms. The mass spectrometer is able to analyze the atoms that make up methane. Scientists can tell where compounds such as methane stemmed from because they have a specific chemical makeup that is unique to the location. The results from the mass spectrometer showed that the methane came from the base of the glacier.

The meltwaters from Katla release up to 45 short tons (41 metric tons) of methane every day during the summer. It would take more than 136,000 belching cows to create that much methane.[6]

Wynn says glaciers provide the perfect environment for methane production: microorganisms, organic matter, water, and low oxygen levels. And because glaciers have a thick layer of ice, the methane gets trapped until the ice melts. Even though Katla wasn't directly responsible for producing the methane, Wynn says that it's still likely playing a role. The warmth from the volcanic activity acts like an incubator. He explains, "It is providing the conditions that allow the microbes to thrive and release methane into the surrounding meltwaters."[5]

The activity at Katla is speeding up more than the methane production. It also melts the ice. The heat from the magma melts the ice from below. Scientists are not sure yet how much this potential rise in methane emissions will affect the climate. There may be a small rise in temperature. However, when the ice is melted, there is no way for more methane to

be created. Scientists say it's possible that the change in temperature could be limited and eventually reverse naturally.

EXTRATROPICAL ERUPTIONS

Volcanoes exist all over the world. When volcanoes erupt, they spew ash, gas, and rocks high into the atmosphere. Large eruptions cause a phenomenon called surface cooling. Surface cooling is the process by which sulfur gases from the volcano in the atmosphere block solar rays. This causes the planet to cool.

Generally, when people think of climate change, they think of the planet warming because of greenhouse gas emissions. Even with the extreme temperatures recorded around the globe in recent years, Earth is actually cooler than it could be. One phenomenon that contributes to this is volcanic eruptions. Scientists estimate that volcanic eruptions

Glacial meltwater feeds this Icelandic waterfall.

between 2000 and 2014 accounted for a surface cooling of 0.09 to 0.22 degrees Fahrenheit (0.05 to 0.12°C).[7]

Volcanoes near the equator are called tropical volcanoes. When a volcano is far from the equator, it's called extratropical. Extratropical eruptions are often smaller than tropical eruptions. Because of this, scientists once concluded that extratropical eruptions had less of an effect on the planet's climate.

Explosive eruptions often throw debris into the lower stratosphere. Gases and ash fly up six to nine miles (10 to 15 km) to reach this part of the atmosphere.[8] This material can linger in the atmosphere for days or even months before dissipating. Some previous evidence had suggested that gases from tropical eruptions lingered longer and had a stronger climatic effect than sulfur from extratropical eruptions. But researchers wanted to verify this link.

To find out whether extratropical eruptions have the same climatic effects as tropical eruptions, researchers turned to an unlikely source: trees. In 2019, Matthew Toohey with GEOMAR Helmholtz Centre for Ocean Research Kiel in Germany led a group of researchers from around the world. They used tree rings to determine periods of cooling and heating. Scientists used drills to take small cylindrical samples of the trees. The tree rings gave scientists a year-by-year catalog of climate conditions by preserving a record of how well or poorly the tree grew each year. The rings dated from 750 CE to the present day.[9] Then researchers compared the trees' growth patterns with known dates of extratropical and tropical eruptions.

Scientists are learning to find clues about the past and future effects of volcanoes all around us, even in trees.

Their results showed that extratropical eruptions had a bigger impact on surface cooling than tropical eruptions of the same volume. Even relatively weak eruptions in extratropical locations had an effect on surface cooling. This cooling can last for months or years.

If volcanoes are responsible for both heating and cooling the planet, do they just cancel out? It's not that simple, unfortunately. A group of scientists tried to find out if they could recreate the surface cooling caused by a large eruption in order to slow the process of climate change. The group used computer simulations in 2018 to see what the possible outcomes would be if an artificial surface cooling were to occur. The simulations worked—the planet cooled. But there was a cost. Some areas got a lot of rain. But others were stuck in a drought. It was unlikely scientists could safely cool the planet using the same techniques as a volcano.

Volcanology is still a new science. Volcanologists are always learning new things about how Earth formed and how life came to inhabit the planet. These scientists use techniques from many scientific disciplines. Volcanologists believe that volcanoes hold the secrets to understanding how life began and how

LIFE ON SATURN'S MOON?

NASA's *Cassini* spacecraft orbited Saturn and its moons. For four years, *Cassini* collected microscopic rocks in Saturn's system. Teams at NASA analyzed these samples. In 2015, they published their findings. *Cassini* found evidence that Saturn's moon Enceladus has hydrothermal activity. The hydrothermal activity resembled the same activity that occurs at the bottom of Earth's oceans. Scientists speculate microorganisms may live near the hydrothermal vents on Enceladus in the same way they do near Earth's hydrothermal vents. With more missions, scientists may confirm for the first time the presence of life beyond Earth.

life continues to transform. Volcanology has even reached space. As our understanding of volcanoes grows, so will our understanding of Earth itself.

EYJAFJALLAJÖKULL ERUPTION IN 2010

The Icelandic volcano Eyjafjallajökull erupted in 2010. Its clouds of ash stopped air traffic for five days. Twenty countries closed their airspace to commercial flights. Thousands of flights were delayed or canceled. Researchers estimate that as many as ten million people were affected by the disruption.[10]

However, the eruption wasn't very big. Its ash plume reached no more than 5.6 miles (8 km) high.[11] The eruption spanned several months. The Icelandic volcano wasn't big enough to cause surface cooling. What made the eruption so troublesome was the amount of time its ash stayed in the air and the direction of the winds. The ash didn't spread out. Instead, it floated along in a stream across Europe.

ESSENTIAL FACTS

SIGNIFICANT EVENTS

- In 1879, the United States Geological Survey formed. The USGS monitors volcanic activity across the United States.

- In 1912, Thomas Jaggar created the Hawaiian Volcano Observatory. The HVO is overseen by the USGS.

- After the 1980 eruption of Mount Saint Helens in Washington, volcanology became a much larger area of study.

- Technological advances in the 2000s allowed volcanologists to understand the inner structure of volcanoes and how the chemical makeup of the lava affects the eruption style of volcanoes.

KEY PLAYERS

- Jeffrey Marlow is a geobiologist best known for his work studying microorganisms in extreme environments such as that at Mount Marum's lava lake.

- Hannah Shamloo was one of the lead researchers on the 2017 study of Yellowstone's caldera. The study found evidence that once the chamber began filling with magma, the timeline to eruption was much faster than scientists originally thought.

- Rebecca Burns found evidence that glacial meltwater holds higher levels of methane than surrounding sources, concluding that volcanic activity combined with glaciers creates an environment in which methane production is elevated.

- Guilherme Gualda led a group to New Zealand to study supereruptions, finding that multiple smaller eruptions in a short time helped avoid huge eruptions because the magma settled closer to the surface. This allowed the lava to escape easily instead of building up.

▸ Ingo Sonder specializes in physical volcanology. In his 2018 experiment mixing water and lava, Sonder proved that the presence of water had the ability to produce highly explosive volcanic eruptions.

IMPACT ON SCIENCE

Volcanology has broadened our understanding not only of eruptions but also of how Earth formed. Scientists are becoming more and more sure that Earth's oceans formed thanks to volcanic activity four billion years ago. By studying where microorganisms are capable of thriving, scientists are able to form theories about where life may exist beyond Earth. Volcanology is still a young science, but it has given researchers many answers to the inner workings of our planet.

QUOTE

"Sometimes, when lava encounters water, you see huge, explosive activity. Other times, there is no explosion, and the lava may just cool down and form some interesting shapes. What we are doing is trying to learn about the conditions that cause the most violent reactions."

—*Ingo Sonder on the lava-water experiment*

GLOSSARY

composition
The makeup of something.

dissipate
To scatter.

dormancy
A period of inactivity.

fluctuation
A change.

geophysicist
A type of geologist who studies the physics of Earth, including volcanology.

imminent
Immediately about to happen.

influx
The flow of substances into an area.

insulated
Protected against extreme temperatures.

photosynthesis
The process used by plants and other organisms to convert sunlight into usable energy.

plume
A column of molten rock that rises from the mantle.

rift zone
An area of cracks along the sides of a volcano where lava erupts.

sediment
Tiny fragments of rock and other particles.

subduction
The process by which one tectonic plate slips underneath another plate.

sulfur
An element that forms a chalky mineral and smells like rotten eggs, often released during volcanic activity.

summit
The top of a volcano or mountain.

turbulent
Irregular or agitated.

ADDITIONAL RESOURCES

SELECTED BIBLIOGRAPHY

"Climate Cooling." *Volcano World, Oregon State University*, n.d., volcano.oregonstate.edu. Accessed 19 June 2019.

"Living Volcanoes." *Nature.* PBS, 20 Feb. 2019.

"What Is a Hydrothermal Vent?" *National Oceanic and Atmospheric Administration*, n.d., oceanservice.noaa.gov. Accessed 19 June 2019.

FURTHER READINGS

Baxter, Roberta. *Seismology: Our Violent Earth*. Abdo, 2015.

Nargi, Lela. *Absolute Expert: Volcanoes*. National Geographic Kids, 2018.

Winchester, Simon. *When the Earth Shakes: Earthquakes, Volcanoes, and Tsunamis*. Viking, 2015.

ONLINE RESOURCES

To learn more about volcanologists, please visit **abdobooklinks.com** or scan this QR code. These links are routinely monitored and updated to provide the most current information available.

MORE INFORMATION

For more information on this subject, contact or visit the following organizations:

HAWAII VOLCANOES NATIONAL PARK
PO Box 52
Hawaii National Park, HI 96718
nps.gov/havo/planyourvisit/craterrimtour_kvc.htm
Hawaii Volcanoes National Park covers the area where Mauna Loa and Kilauea sit. Visitors can hike in this unique environment and see how Hawaii is constantly changing.

MOUNT SAINT HELENS VISITOR CENTER
3029 Spirit Lake Highway
Castle Rock, WA 98611
parks.state.wa.us/245/Mount-St-Helens
The Mount Saint Helens Visitor Center provides visitors with information regarding the 1980 eruption. A permanent exhibit gives visitors a glimpse at the landscape before and during the eruption and the lasting effects the eruption had on wildlife.

YELLOWSTONE NATIONAL PARK
PO Box 168
Yellowstone National Park, WY 82190-0168
nps.gov/yell/planyourvisit/index.htm
Yellowstone was the United States' first national park. The park is known for its hot springs and geysers, and visitors can see firsthand the volcanic activity fueling these hydrothermal features.

SOURCE NOTES

CHAPTER 1. LIFE IN A LAVA LAKE

1. "Living Volcanoes." *Nature*, season 37, episode 11, PBS, 20 Feb. 2019, pbs.org. Accessed 22 July 2019.

2. "How Many Active Volcanoes Are There on Earth?" *USGS*, n.d., usgs.gov. Accessed 22 July 2019.

3. "How Hot Is Lava?" *Volcano World*, n.d., volcano.oregonstate.edu. Accessed 22 July 2019.

4. "Living Volcanoes."

5. Jason Daley. "How Low Can Life Go? New Study Suggests Six Miles Down." *Smithsonian*, 18 Apr. 2017, smithsonianmag.com. Accessed 22 July 2019.

6. "Living Volcanoes."

7. "Vesuvius, Italy." *Volcano World*, n.d., volcano.oregonstate.edu. Accessed 22 July 2019.

8. Doug Stewart. "Resurrecting Pompeii." *Smithsonian*, Feb. 2006, smithsonianmag.com. Accessed 22 July 2019.

CHAPTER 2. DECADES OF DATA

1. Charles Q. Choi. "Tiny Gravity Changes Show Magma's Underground Movements." *LiveScience*, 31 Aug. 2012, livescience.com. Accessed 22 July 2019.

2. Choi, "Tiny Gravity Changes."

3. Choi, "Tiny Gravity Changes."

4. Choi, "Tiny Gravity Changes."

5. "Kilauea Volcano in Hawaii." *Volcanoes*, n.d., volcanoes.org.uk. Accessed 22 July 2019.

6. "Kilauea Volcano in Hawaii."

7. Brett Israel. "Kilauea Volcano's Deadliest Eruption Revealed." *LiveScience*, 6 Dec. 2011, livescience.com. Accessed 22 July 2019.

8. "Volcano Watch: New Insights Gained from Kilauea Volcano's 2018 Summit Collapses." *USGS Volcano Hazards Program, Hawaiian Volcano Observatory*, 9 May 2019, volcanoes.usgs.gov. Accessed 22 July 2019.

9. Becky Oskin. "Volcano's 30-Year Eruption Bursting with Discoveries." *LiveScience*, 3 Jan. 2013, livescience.com. Accessed 22 July 2019.

10. "Volcano." *Encyclopedia Britannica*, 17 July 2019, britannica.com. Accessed 22 July 2019.

CHAPTER 3. SUPERVOLCANOES

1. Heidelberg University. "Traces of Failed Super-Eruption in the Andes." *ScienceDaily*, 2 Aug. 2016, sciencedaily.com. Accessed 22 July 2019.

2. Sarah Pruitt. "Scientists Solve Supervolcano Mystery." *History*, 8 Jan. 2014, history.com. Accessed 22 July 2019.

3. Shannon Hall. "A Surprise from the Supervolcano under Yellowstone." *New York Times*, 10 Oct. 2017, nytimes.com. Accessed 22 July 2019.

4. David Cox. "Would a Supervolcano Eruption Wipe Us Out?" *BBC Future*, 24 July 2017, bbc.com. Accessed 22 July 2019.

5. Charles Q. Choi. "Catastrophic Volcanoes Blamed for Earth's Biggest Extinction." *LiveScience*, 28 Aug. 2015, livescience.com. Accessed 22 July 2019.

6. Hall, "A Surprise from the Supervolcano under Yellowstone."

7. Hall, "A Surprise from the Supervolcano under Yellowstone."

8. Robin George Andrews. "Campi Flegrei Volcano's Ancient Cycle Seems to End in Large Eruption." *New York Times*, 14 Nov. 2018, nytimes.com. Accessed 22 July 2019.

9. "How Hot Is the Yellowstone Hotspot?" *LiveScience*, 12 Sept. 2012, livescience.com. Accessed 22 July 2019.

10. Roberto Moretti, Giuseppe De Natale, and Claudia Troise. "A Geochemical and Geophysical Reappraisal to the Significance of the Recent Unrest at Campi Flegrei Caldera (Southern Italy)." *Geochemistry, Geophysics, Geosystems*, vol. 18, no. 3, Mar. 2017, *AGU100*, agupubs.onlinelibrary.wiley.com. Accessed 22 July 2019.

11. Hannah Osborne. "Campi Flegrei: One of the World's Most Dangerous Supervolcanoes Could Erupt Sooner Than Expected." *Newsweek*, 15 May 2017, newsweek.com. Accessed 22 July 2019.

12. Osborne, "Campi Flegrei."

13. David Cox. "Nasa's Ambitious Plan to Save Earth from a Supervolcano." *BBC Future*, 17 Aug. 2017, bbc.com. Accessed 22 July 2019.

CHAPTER 4. WHAT MAKES A VOLCANO EXPLODE?

1. "Volcanoes/Lava Flows." *National Park Service: Lassen Volcanic National Park, California*, 4 May 2019, nps.gov. Accessed 22 July 2019.

2. Ann Taylor. "The Eruption of Mount St. Helens in 1980." *Atlantic*, 18 May 2015, theatlantic.com. Accessed 22 July 2019.

3. University at Buffalo. "Scientists Brew Lava and Blow It Up to Better Understand Volcanoes." *ScienceDaily*, 10 Dec. 2018, sciencedaily.com. Accessed 22 July 2019.

4. University at Buffalo, "Scientists Brew Lava."

5. "Q&A: What Happens When a Volcano Beneath a Glacier Erupts?" *National Science Foundation*, 11 Oct. 2016, nsf.gov. Accessed 22 July 2019.

CHAPTER 5. RING OF FIRE

1. "Ring of Fire." *National Geographic Resource Library*, 5 Apr. 2019, nationalgeographic.org. Accessed 22 July 2019.

2. Mary Bagley. "Krakatoa Volcano: Facts about 1883 Eruption." *LiveScience*, 14 Sept. 2017, livescience.com. Accessed 22 July 2019.

3. "Mount Pinatubo." *Encyclopedia Britannica*, 3 June 2019, britannica.com. Accessed 22 July 2019.

4. Kathy Svitil. "The Volcanoes of North America." *Savage Earth*, n.d., thirteen.org. Accessed 22 July 2019.

5. "Ring of Fire Full Documentary." *YouTube*, uploaded by Natural World, 28 May 2014, youtube.com. Accessed 22 July 2019.

6. "Ring of Fire Full Documentary."

7. Ella Koscher. "How Volcano 'Voiceprints' Could Help Predict Eruptions." *NBC News*, 19 July 2018, nbcnews.com. Accessed 22 July 2019.

107

8. Becky Oskin. "Volcano's 'Infrasound' Roar Is a Weather Vane." *LiveScience*, 18 Apr. 2013, livescience.com. Accessed 22 July 2019.

9. Oskin, "Volcano's 'Infrasound' Roar Is a Weather Vane."

10. Koscher, "How Volcano 'Voiceprints' Could Help Predict Eruptions."

11. "Bat Echolocation." *Maryland Department of Natural Resources*, n.d., dnr.maryland.gov. Accessed 22 July 2019.

12. J. B. Johnson et al. "Infrasound Tornillos Produced by Volcán Cotopaxi's Deep Crater." *Geophysical Research Letters*, vol. 45, 13 June 2018, *AGU100*, agupubs.onlinelibrary.wiley.com. Accessed 22 July 2019.

13. "Determining the Depth of an Earthquake." *USGS*, n.d., usgs.gov. Accessed 22 July 2019.

CHAPTER 6. VOLCANOES AND OCEANS

1. "Comets or Volcanoes?" *Deepstuff*, 6 Mar. 2017, deepstuff.org. Accessed 22 July 2019.

2. "Comets or Volcanoes?"

3. Peter Ulmer and Volkmar Trommsdorff. "Serpentine Stability to Mantle Depths and Subduction-Related Magnetism." *Science*, 12 May 1995, science.sciencemag.org. Accessed 22 July 2019.

4. Nola Taylor Redd. "How Old Is Earth?" *Space*, 7 Feb. 2019, space.com. Accessed 22 July 2019.

5. "Why Do We Have Oceans?" *National Ocean Service*, n.d., oceanservice.noaa.gov. Accessed 22 July 2019.

6. Brandon Schmandt et al. "Dehydration Melting at the Top of the Lower Mantle." *Science*, 13 June 2014, science.sciencemag.org. Accessed 22 July 2019.

7. Schmandt et al., "Dehydration Melting."

8. "Submarine Volcanoes." *Volcano World*, n.d., volcano.oregonstate.edu. Accessed 22 July 2019.

9. "Seamount." *Encyclopedia Britannica*, n.d., britannica.com. Accessed 22 July 2019.

10. "Submarine Volcanoes."

11. Oliver Milman. "Huge and Ancient Underwater Volcanoes Discovered Off Coast of Sydney." *Guardian*, 12 July 2015, theguardian.com. Accessed 22 July 2019.

12. "What Is a Hydrothermal Vent?" *National Ocean Service*, n.d., oceanservice.noaa.gov. Accessed 22 July 2019.

13. "Deep Sea Hydrothermal Vents." *National Geographic Resource Library*, 21 Mar. 2013, nationalgeographic.org. Accessed 22 July 2019.

CHAPTER 7. PREDICTING ERUPTIONS

1. University of Bristol. "Satellite View of Volcanoes Finds the Link between Ground Deformation and Eruption." *ScienceDaily*, 3 Apr. 2014, sciencedaily.com. Accessed 22 July 2019.

2. University of Bristol, "Satellite View of Volcanoes."

3. University of Bristol, "Satellite View of Volcanoes."

4. University of Bristol. "Satellite Images Reveal Interconnected Plumbing System That Caused Bali Volcano to Erupt." *ScienceDaily*, 14 Feb. 2019, sciencedaily.com. Accessed 22 July 2019.

5. University of Bristol, "Satellite Images Reveal Interconnected Plumbing System."

6. University of Bristol, "Satellite Images Reveal Interconnected Plumbing System."

7. Vanderbilt University. "Volcano Researcher Learns How Earth Builds Supereruption-Feeding Magma Systems." *ScienceDaily*, 10 Oct. 2018, sciencedaily.com. Accessed 22 July 2019.

8. Vanderbilt University, "Volcano Researcher."

9. University of New Hampshire. "Eruption Clues: Researchers Create Snapshot of Volcano Plumbing." *ScienceDaily*, 29 Nov. 2017, sciencedaily.com. Accessed 22 July 2019.

10. University of Queensland. "'Amazing Snapshots' Plumb Volcanic Depths." *ScienceDaily*, 28 Feb. 2019, sciencedaily.com. Accessed 22 July 2019.

11. University of Queensland, "'Amazing Snapshots.'"

CHAPTER 8. ERUPTIONS AND EARTH'S CLIMATE

1. Uppsala University. "Large Volcanic Eruption in Scotland May Have Contributed to Prehistoric Global Warming." *ScienceDaily*, 24 Jan. 2019, sciencedaily.com. Accessed 22 July 2019.

2. Uppsala University, "Large Volcanic Eruption in Scotland."

3. Lancaster University. "Volcanoes and Glaciers Combine as Powerful Methane Producers." *ScienceDaily*, 20 Nov. 2018, sciencedaily.com. Accessed 22 July 2019.

4. Robin McKie. "Scientists Discover 91 Volcanoes Below Antarctic Ice Sheet." *Guardian*, 12 Aug. 2017, theguardian.com. Accessed 22 July 2019.

5. Lancaster University, "Volcanoes and Glaciers Combine."

6. Lancaster University, "Volcanoes and Glaciers Combine."

7. Dana Nuccitelli. "Volcanoes May Be Responsible for Most of the Global Surface Warming Slowdown." *Guardian*, 3 Dec. 2014, theguardian.com. Accessed 22 July 2019.

8. Helmholtz Centre for Ocean Research Kiel. "Extratropical Volcanoes Influence Climate More Than Assumed." *ScienceDaily*, 28 Jan. 2019, sciencedaily.com. Accessed 22 July 2019.

9. Helmholtz Centre, "Extratropical Volcanoes."

10. "Seven Years Ago Today: Eruption in Eyjafjallajökull, the Volcano with the Un-pronouncable Name." *Iceland Magazine*, 14 Apr. 2017, icelandmag.is. Accessed 22 July 2019.

11. "Monitoring the Eyjafjallajökull Eruption." *Climate.gov*, 22 Apr. 2010, climate.gov. Accessed 22 July 2019.

INDEX

ABOUT THE AUTHOR

Martha London works as a writer and educator. She lives in Saint Paul, Minnesota, and in her spare time she enjoys hiking.